All illustrations in this book were made by Adya Jha
Back Cover Photo by Pixabay from Pexels (Canva Stock Photos)

Title: Carl Sagan: the Boy Who Looked to the Stars

Author: Adya Jha

ISBN: 9781086059908

Contact the author at her blog- adyajha.com

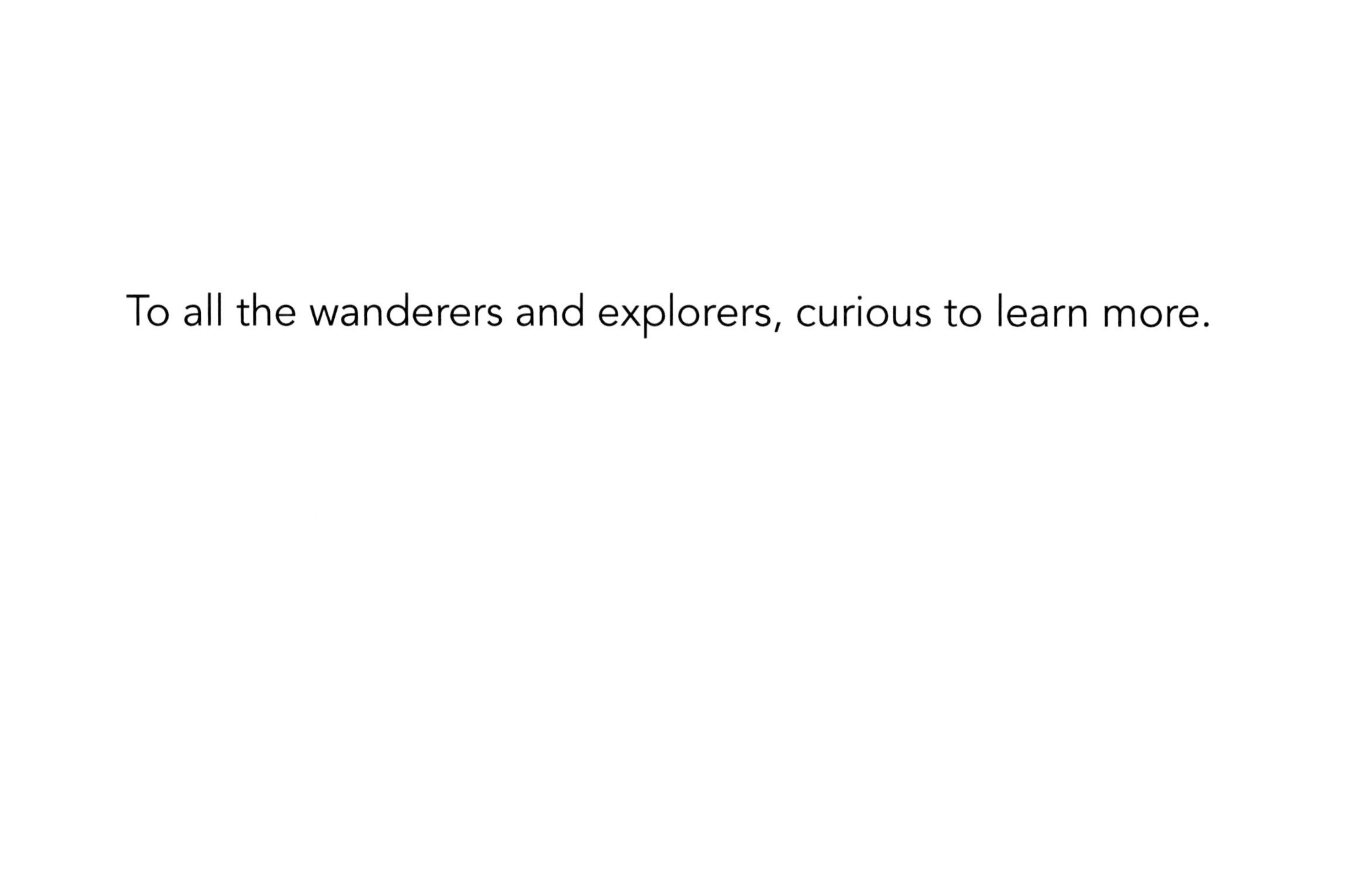

To all the wanderers and explorers, curious to learn more.

Right now, hurtling through space at 38,610 miles per hour is Voyager 1. Voyager 1 was made to explore our **solar system** and beyond. Voyager 1 carries messages for **aliens** to find.

Collected on Voyager 1 are several songs from around the world, greetings in other languages, images of a woman and a man, and directions to Earth. They are compiled on a Golden Record. The messages on Voyager 1 were the idea of the famous **astronomer**, Carl Sagan.

THE

When he was a child, Carl had big dreams about what lay in the **cosmos**. He had questions about the stars and the universe. He had many questions and he needed answers. On his mother's suggestion, he went to the library and explored the world of books. He didn't just want to learn about the universe, he wanted to learn the very nature of the universe. He found that space was breathtaking.

He visited museums and the **Hayden Planetarium** in New York, and was inspired by the exhibits. Carl loved to work with numbers. He was curious about his surroundings and always wondered how things worked.

1939 NEW YORK
WORLD'S FAIR

Were we all alone in the universe? Carl continued to have questions. His parents took him to the 1939 New York World's Fair. Carl saw things that were unimaginable at that time, such as robots and air conditioning. Carl knew from then on that he wanted to be a **scientist**.

THE WAR OF THE WORLD

He would spend his days with his nose in a book, reading about things like aliens and the future. Carl also enjoyed **science fiction**; he loved to read short stories. Science fiction may have inspired him to be interested in **extraterrestrial life**. Perhaps this was why he was excited to send messages on a **spacecraft** where aliens might find them.

ASTRONOMY
CHEMISTRY
ADVANCED PHYSICS
MATHEMATICS

At sixteen, Carl went to the University of Chicago after finishing at Rahway High School. He was very smart and focused on **physics**, the study of matter and motion. When Carl was unhappy with his grades, he only took this as a sign to work harder. He worked late into the night, studying.

Carl also loved other types of science too. He loved biology, studying how life works, almost as much as astronomy. Carl spent his summers writing letters to famous scientists like H.J. Muller, Harold Urey and Gerard Kuiper.

RULES OF
THE
LAB
-SAFETY
GOGGLES

He had the opportunity to work with them in their **labs**. Harold Urey was a chemist, a scientist who studies what matter is made of. Carl Sagan worked with him in a class about the beginning of life. Harold Urey worked on the **atom bomb**; Carl would listen to his lectures and get inspired.

Carl didn't stop there. He became a brilliant scientist, and discovered many new things. He was known for not just thinking about Earth, but taking in the entire universe. He was a professor at Cornell University, and he took great care in his students' education. In the 1970s, Carl Sagan headed the team that worked on the Golden Record that was placed on the Voyager spacecraft.

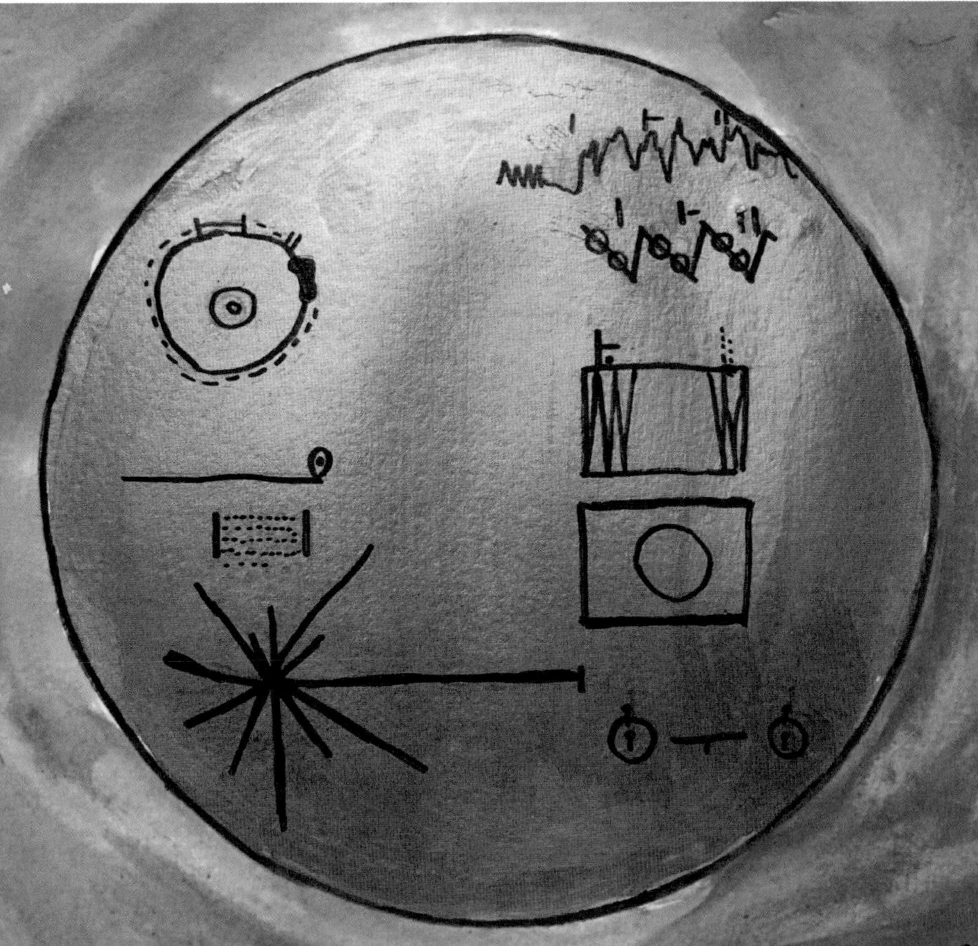

THE GOLDEN RECORD - COVER WITH ILLUSTRATIONS

Carl Sagan and his team collected several glimpses into life from all around planet Earth. If any extraterrestrial life was to ever find Voyager, they would receive a window into the human race's language, culture, and world.

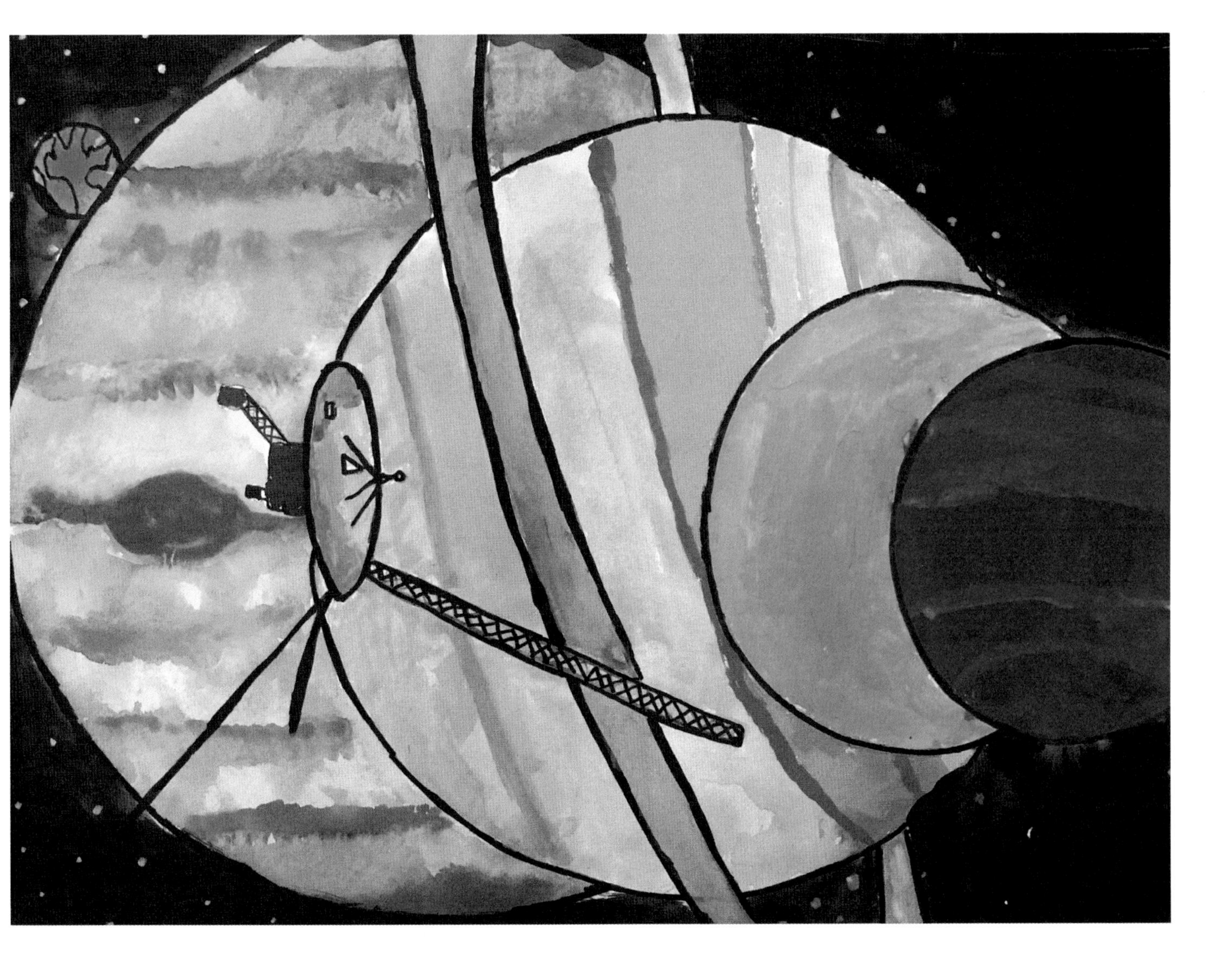

When Voyager 1 launched in the 1970s, Carl Sagan's Golden Record took a journey across the solar system. Voyager went on a "Grand Tour." Voyager visited **planets** and observed their **moons**. Voyager took decades to complete this tour. The information it gathered proved extremely useful to scientists on Earth. Voyager 1 crossed into **interstellar space** in 2012. It is predicted that Voyager 1 will take around 40,000 years to reach a nearby star, Gliese 445. In 200 million years, Voyager will complete its first roundabout of our galaxy.

After Voyager, Carl Sagan went on to achieve even greater things. He thought that others should understand science too. He felt that it was necessary for the everyday person to know about space. He hosted the television series Cosmos. It was an account of the beauty of the universe.

Not everyone believed in his work, however. In 1983, Carl Sagan presented a theory about a nuclear winter. A nuclear winter is when the Earth becomes very cold, as a result of a nuclear war. This causes many plants and animals to die off. Other scientists thought that Carl was crazy. Although they thought he was being a lunatic, Carl never backed down. He wouldn't let them destroy what he believed in.

"a work of art!"
CARL SAGAN
COSMOS
"amazing!"
CARL SAGAN
A NOVEL
CONTACT
CONTACT-CARL SAGAN
"stupendous!"
"beautiful!"
CARL SAGAN
THE DRAGONS OF EDEN
speculation on the evolution of human intelligence
WINNER OF THE PULITZER PRIZE
BILLIONS & BILLIONS
CARL SAGAN
IN THE DARK
"the best!"

Carl wrote many books, such as *Pale Blue Dot* and A Demon Haunted World. He wrote one science fiction book, *Contact*. His book, *The Dragons of Eden*, won the Pulitzer Prize. One of his most celebrated books is *Cosmos*. In *Cosmos*, he used beautiful writing to show us how wonderful the universe is. His writing is known for being gorgeous and poignant.

Carl Sagan's legacy is still present today. He has shown us the depths of space in *Cosmos*, and has shared with us the mysteries of the universe in his books. Carl Sagan has taught us what it is to believe, and make one's beliefs come true.

Voyager 1 is still silently moving through the depths of the cosmos, an unfathomable idea. A piece of earth is slowly moving away, a record of what Carl Sagan thought beautiful.

Carl Sagan

Carl Sagan (1934-1996), was born into a relatively lower class family in Brooklyn, New York City. He was an exceptionally inquisitive child, and he was hard-working. He attended Rahway High School in New Jersey, and went on to study at the University of Chicago in Chicago, Illinois. Sagan was a science popularizer, and through his television show, *Cosmos*, he became one of the most well known scientists of the 20th century. He led the Golden Record team in the Voyager mission. The Golden Records are still on Voyager 1 and Voyager 2, and the two spacecraft recently crossed into interstellar space. Aside from heading the Golden Record, Sagan published several books. He was a political advocate, as well as the Co-Founder of SETI, the Search for Extraterrestrial-Life Institute. Sagan died at 62, after suffering from pneumonia. Carl Sagan will be remembered for his contributions to science, as well as his commitment to making science available to the general public.

Author's Note

Ever since I was a kid (and I still am!), I have admired the stars. I first watched Neil deGrasse Tyson's *Cosmos* when it came out, in 2014. I was seven at the time. I was immediately enraptured by the story of the universe. I dug a little deeper, and online, I found a few clips of Carl Sagan's original *Cosmos* television show. A few years later, I lost myself in the mystical prose of his bestselling book of the same name, *Cosmos*. Carl Sagan became one of my main inspirations, and in fifth grade, I started researching his life and work. I decided I wanted to write a book about this phenomenal man, and my project combined my two greatest interests: writing and science. The process of writing this book took two years, and I made several mistakes in getting to the final product. I learned to never give up and stay persistent. I have come away from this book with a far more comprehensive knowledge of Carl Sagan, and a drive to inspire more little kids to pursue science, just as Carl Sagan once inspired me.

A Note on the Illustrations

I made these illustrations over a course of five months. I used watercolor and helpful online tutorials to achieve the illustrations you see throughout this book. Before illustrating this book, I had never really made any art seriously, and did not draw frequently. While the illustrations in *Carl Sagan- the Boy Who Looked to the Stars*, are not perfect, I did my best to capture the overall essence of his illustrious life.

Acknowledgements

I would like to recognize Mr. Michael Lemonick, science writer, who sat down with me and discussed Carl Sagan. I would also like to thank Dr. Ingrid Ockert, Princeton University science historian, for directing me to the right resources. The inspiration for this book would also not have been possible without Dr. Akhilesh Singh, physicist, who taught me that physics could be fun. I would like to thank Mrs. Bev Gallagher for helping me with my writing and encouraging me to pursue this project. Finally, I would like to thank my two sisters, as well as my parents for guiding me along these two years I have been writing this book.

Glossary

Aliens- A being from another world

Asteroid- A minor terrestrial body orbiting the Sun

Asteroid Belt- A ring of asteroids circling our Sun

Astronomer- A scientist who studies space

Astronomy- A branch of science studying space

Atom Bomb- A bomb set off by the United States to end war with Japan

Cosmos- The universe and space

Extraterrestrial Life- Life existing outside of Earth

Hayden Planetarium- A theatre where people can see and be educated of the night sky

Interstellar Space- The space between stars

Labs- Places where one can experiment and study science

Moons- Celestial bodies that orbit planets

Nuclear Winter- The result of a nuclear war-- the Earth becomes very cold, and many plants and animals cannot survive

Planets- Celestial bodies (can be terrestrial or made of gas) that orbit stars

Science Fiction- A genre where stories are fictional, but have a scientific part

Scientist- One who studies, observes and creates experiments about the world

Solar system- A family of planets circling a star

Spacecraft- A vehicle deployed to study space

Theory- An explanation for how and why things happen

Universe- The extent of space and time

Bibliography

1. Davidson, Keay. Carl Sagan: a Life. J. Wiley, 2000.

2. NASA. [Online]. Available: https://voyager.jpl.nasa.gov/. [Accessed: 12-Feb-2019]

3. NASA. [Online]. Available: https://voyager.jpl.nasa.gov/. [Accessed: 12-Feb-2019]

"Sagan's Youth and the Progressive Promise of Space - Library of Congress Information Bulletin. [Online]. Available: https://www.loc.gov/collections/finding-our-place-in-the-cosmos-with-carl-sagan/articles-and-essays/carl-sagan-and-the-tradition-of-science/sagans-youth-and-the-progressive-promise-of-space

4. "When Carl Sagan Warned the World About Nuclear Winter," Smithsonian.com, 15-Nov-2017. [Online]. Available: https://www.smithsonianmag.com/science-nature/when-carl-sagan-warned-world-about-nuclear-winter-180967198/. [Accessed: 12-Feb-2019]

5. "Who Was Carl Sagan?" National Geographic, National Geographic Society, 17 Mar. 2014, news.nationalgeographic.com/news/2014/03/140316-carl-sagan-science-galaxies-space/.

About the Author

Adya Jha is a 12 year old writer from Princeton, New Jersey. She enjoys writing memoirs, non-fiction, fiction and biographies. She first shared her writing with the world at eight, when she started her blog. For the past two years, Adya has published a newspaper called 54 Tribune for her family and friends. *Carl Sagan- the Boy who Looked to the Stars* is her first book. Adya loves traveling, exploring new places, and food. Adya is inspired by scientists such as Carl Sagan, Stephen Hawking and Annie Jump Cannon. You can visit her at her blog at adyajha.com.

Carl Sagan- From Skeeze on Pixabay

13310376R00021

Made in the
USA
Monee, IL